WOODY WOODPECKER

AMAZING NIGHT ANIMALS

BETH HARWOOD

CONTENTS

Project Manager: Belinda Weber
Art Editor: Eljay Yildirim
Designer: Dean Price
DTP Manager: Nicky Studdart
Production Controller: Jo Blackmore

KINGFISHER/UNIVERSAL
Kingfisher Publications Plc
New Penderel House
283-288 High Holborn
London WC1V 7HZ
www.kingfisherpub.com

First published by Kingfisher Publications Plc 2002

10 9 8 7 6 5 4 3 2 1

1TR/1201/TWP/MAR(MAR)/130SINAR

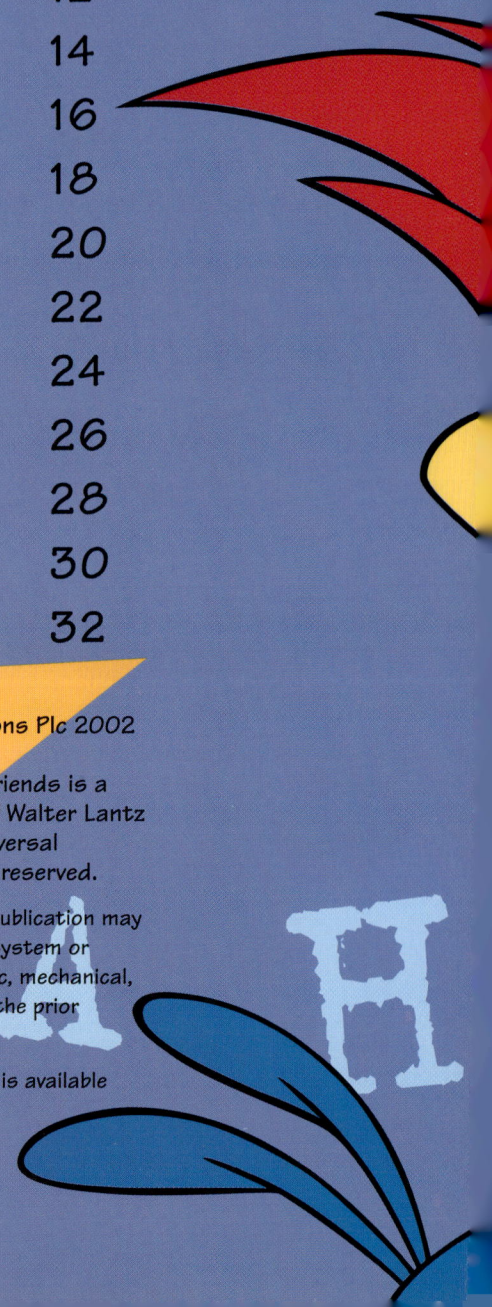

Woody Woodpecker™ is a day bird. He wakes at dawn, works during the day, and sleeps through the night. But the sounds and calls of the creatures that are active during the night often wake him. His friends tell him that many animals are awake at night. They are called nocturnal animals.

Some live in hot countries, including Africa or Australia, where it is too warm to hunt during the day. Some creatures are shy and prefer to come out at night when nobody can see them, while others like to hunt in the dark when their prey cannot see them coming. Woody™ decides to stay up late and find out what's going on!

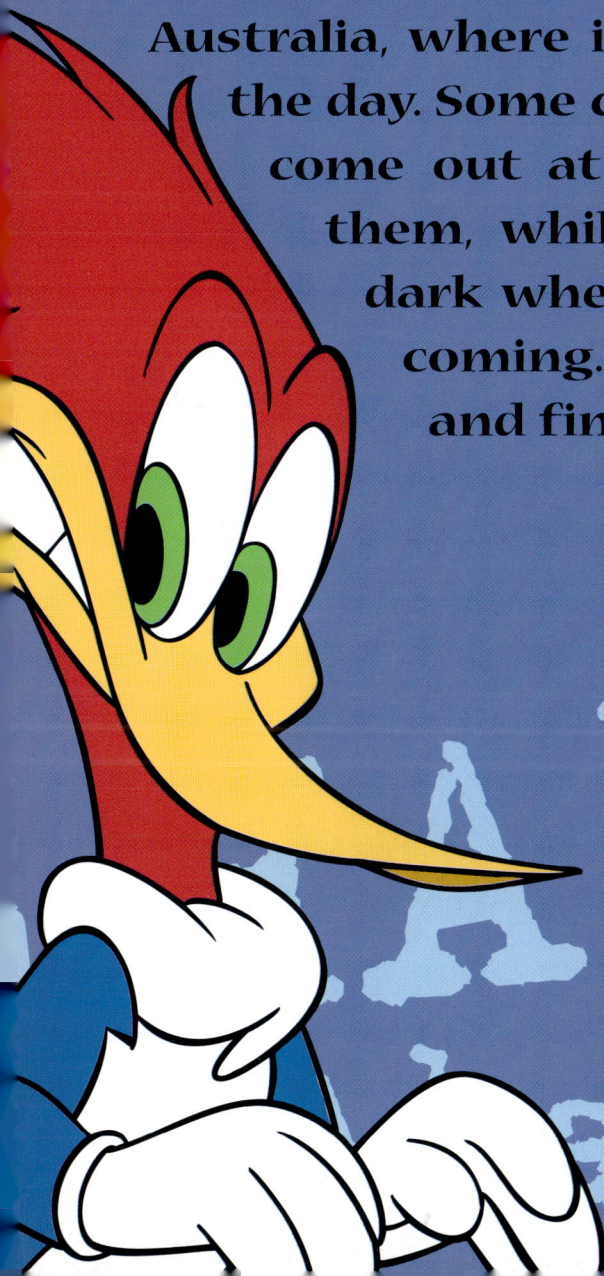

Ha Ha ha ha ha Ha ha ha ha

On the town

Woody™ is taking a journey into town, to sample the bustling nightlife and meet some of the nocturnal creatures that live there. These creatures often have to use keen survival instincts to find a meal. Some animals rummage through garbage cans, while others manage to find their way into homes and feed on humans' leftovers!

WRIGGLY WOODLICE
Woodlice are related to crabs and lobsters. They live underground and come out to hunt at night.

CAT'S WHISKERS
Cats can hunt in pitch-black conditions, using their super-sensitive whiskers to find their way.

LIVING ROUGH
Rats are able to live in harsh, dirty conditions. Their keen hearing and sense of smell help them to find food.

NIGHT BITES

Bedbugs make their homes in mattresses and in any hidden places near a warm bed. They come out at night to feed on the blood of sleeping mammals, including humans!

A SWEET TOOTH

Silverfish appear in bathrooms or kitchens at night. They eat sugar and starch, and even feed on the gluey bindings of old books!

SLIMY SLUGS

Slugs need to keep their bodies wet. They only come out at night when it is damp.

DEADLY TRAP

House spiders spin their webs in dark places to trap insects. They cover their prey in silk before eating them.

Garden gang

Woody™ knows that lots of insects live outside in the garden. Creatures like cockroaches are sensitive to light, and only hunt at night, when they feed on dead creatures or fruits that have fallen from the trees. Woody™ is amazed to find that in the city, these highly adaptable creatures will feed on anything they can find – even shoe polish!

GETTING A GRIP

Earwigs live in dark, hidden places and hunt at night. They feed on fruit, plants and other insects, which they grip with their pincers.

NIGHT FEEDERS

Snails have moist bodies, protected by shells. They hunt at night when the air is fresh, feeding on plants. Snails live underground during the day.

Battling beetles

Stag beetles feed on the sap of trees. They look fierce, but are really quite timid, flying at night to avoid predators. When looking for a mate, male stag beetles fight battles at dusk, grappling their rivals with their jaws and sometimes lifting them off their feet. The fights look dangerous, but they are not because the beetles' jaws cannot close completely. The loser walks away after the fight.

EVIL WEEVILS

The giraffe weevil is so called because of its long neck. The grain weevil feeds on grain and lays its eggs in the kernels. Both are more active at night.

Grain weevil

Giraffe weevil

Wacky World

Some people believe that the ticking of the deathwatch beetle signals death. But the ticking simply shows that a beetle is about to emerge from its home.

Country cousins

Woody™ thinks that creatures who live in the countryside should find it easy to hunt at night. There are plenty of places to find food – in the trees, among the plants, even in the water. Many creatures are camouflaged for protection in the dim light.

Toad

NIGHT AIR

Timid toads prefer to hunt for their food at night when the air is damp. Frogs use their big eyes to help them find their prey, lunging at small insects.

Frog

Wacky World

Frogs' and toads' skin is covered with a poisonous slime that makes it taste bad. Some, like the golden poison-arrow frog from Colombia, are so poisonous that they can kill much larger animals, including people.

NIGHT VISION

Field mice have bulging eyes and sensitive whiskers that help them find their way in the dark. They are good climbers and can jump well.

Summer nights

Warm, summer nights are great for creepy crawlies. Many flying bugs, including moths, use the Moon and stars to help them find their way. But they have to watch out for predators, such as bats, which are also active in the cool of the night.

DANGER!
Moths fly at night. They often fly too close to bright lights and get burnt.

BRIGHT SPARKS
Glow-worms use their own built-in torch to communicate at night.

FLASHY BEETLES
Fireflies are beetles that flash light from their bellies to attract a mate.

Country tracks

Exploring further into the countryside, Woody™ thinks that if he cannot see other creatures, they cannot see him! But many nocturnal creatures can communicate at night. Nightingales, for example, are unusual because they sing by night as well as by day. They are said to be the finest singers of all the birds.

PRICKLY CUSTOMERS

Hedgehogs sleep during the day camouflaged and protected by their spines. At night, they search for food, using their senses of smell and hearing.

Playing with fire

Fireflies are members of the beetle family and are related to glow-worms. Both males and females flash light signals to potential mates, but females do not fly. The light, beamed from the underbelly, also warns predators that the firefly is not good to eat.

OFF THE RAILS

Railway worms are the larvae of a rare South American firefly. Their tiny body lights and big red 'headlamp' look like railway signals.

NIGHT LIGHT

Female glow-worms glow to attract males. Larvae feed until they are adult, then they do not eat at all!

Wacky World

Fireflies are able to control their glow. The female firefly glows roughly every two seconds, while males flash once every five seconds. One North American species even tell lies! The female flashes the signal of another species, and when a male comes near, she eats him!

11

Wild woods

Woody™ is used to being in the woods during the day, but at night, he can see and hear creatures he has never come across before! Many creatures that live in the woods are dark coloured so that they blend in with their surroundings. Woody™ thinks he should hide, in case they see him!

CRAFTY FOX

Foxes are adaptable animals that will eat almost anything. They hunt at night, feeding on smaller animals such as chickens. They attack without warning!

FAMILY HOMES

Badgers hunt at night, snuffling up fruits and small creatures. They are creatures of habit and often use the same paths when searching for food.

NIGHT HUNTERS

Raccoons hunt by night in the town or country when they feel less threatened. They are called 'bandits of the night' because of their cute black and white 'masked' faces.

PACK OF WOLVES

Wolves belong to the same family as dogs. They live in family groups or packs made up of a pair of adults and several generations of their young. They often travel at night.

Bats galore

Bats are the only mammals that can fly. There are 900 different types of bat, ranging from fruit bats to the blood-sucking vampire bats. Most bats cannot see very well so hunt by echo-location. To do this, a bat squeaks and if there is prey nearby, the bat hears an echo and can follow it to find the prey.

NIGHTMARE!

Vampire bats feed solely on blood, usually from cattle or horses. They have razor-sharp teeth that can slice into their sleeping prey.

OLD MEN

Woody™ thinks 'Old Man' bats look funny. They have wizened faces and white 'beards', hence the name. Their wrinkles help them make their squeaking sounds.

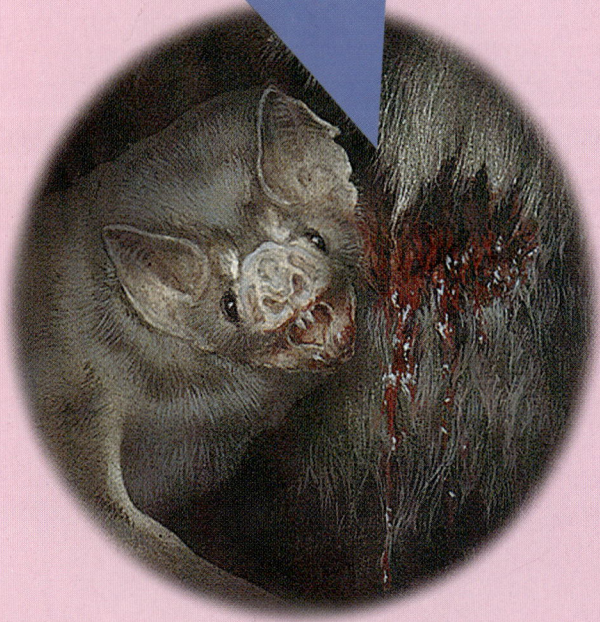

14

FRUITY TASTES

Fruit bats are the largest bats. They have excellent eyesight and a strong sense of smell. They are also called flying foxes because of the shape of their faces.

FAST FLIERS

Mexican free-tailed bats feed on hundreds of tonnes of insects in one night. They live in colonies of up to 20 million other bats, but they can pick out their own young in the huge crowd by smell and cry alone!

FEATHERED FUN!
What is the first thing that bats learn at school?

The alphabat.

FEATHERED FUN!
What do you call a bat in a bell tower?

A dingbat.

In the sea

Woody™ takes a dive beneath the sea, where he meets strange animals like the bat starfish, which has five webbed arms, and prickly flesh that protects it from predators. They stay in dark, damp places to stop their bodies from drying out.

Wacky World

Starfish are really amazing creatures. They can move in any direction without turning. Some have less than four arms; others have as many as nine! Amazingly, if a bat starfish loses an arm during a fight, it will miraculously grow back!

BLENDING IN

Sea horses have translucent flaps on their backs, camouflaging them among the seaweed. Female sea horses compete with each other for male partners.

DEADLY JOKE

Sea anemones fool prey because they look more like flowers than animals. They stay below the water to keep their skin moist.

Swimming ghosts

Moon jellyfish swim near the surface of shallow waters. They are so-called because of their ghostly resemblance to the Moon. They cover their prey in slimy mucus before eating them.

HIDE AND SEEK
Spiny lobsters look like the coral they hide in during the day. At night, they come out to hunt for food. Although they can swim, spiny lobsters prefer to walk along the seabed.

Dastardly deep

Woody™ wonders why many animals in the deep, dark sea look so odd! Anglerfish, for example, lure their prey with the strange glowing rod on their heads. Dragonfish look like Chinese dragons, with light glowing under their skin, while gulper eels swim with their huge mouths open, scooping up any fish they catch.

Anglerfish

Wacky World

The largest ray in the world is the manta ray, which can measure nearly 7 metres across. With its huge wing-like fins, the ray flaps through the sea like an underwater bird. In the past, these fish inspired stories of sea monsters, but they cannot hurt people.

FLOORED!
Most rays live on the seabed, lying camouflaged against the sand, rising to feed on plankton. Their dark colouring means they can hide on the sea floor.

Giant of the deep

Giant clams have the largest shells on earth. As they grow, giant clams become wedged in gaps in the corals, protecting them from the waves. Their outer shell is plain, blending in with rocks and coral reefs, but their colourful inner shell lures potential prey.

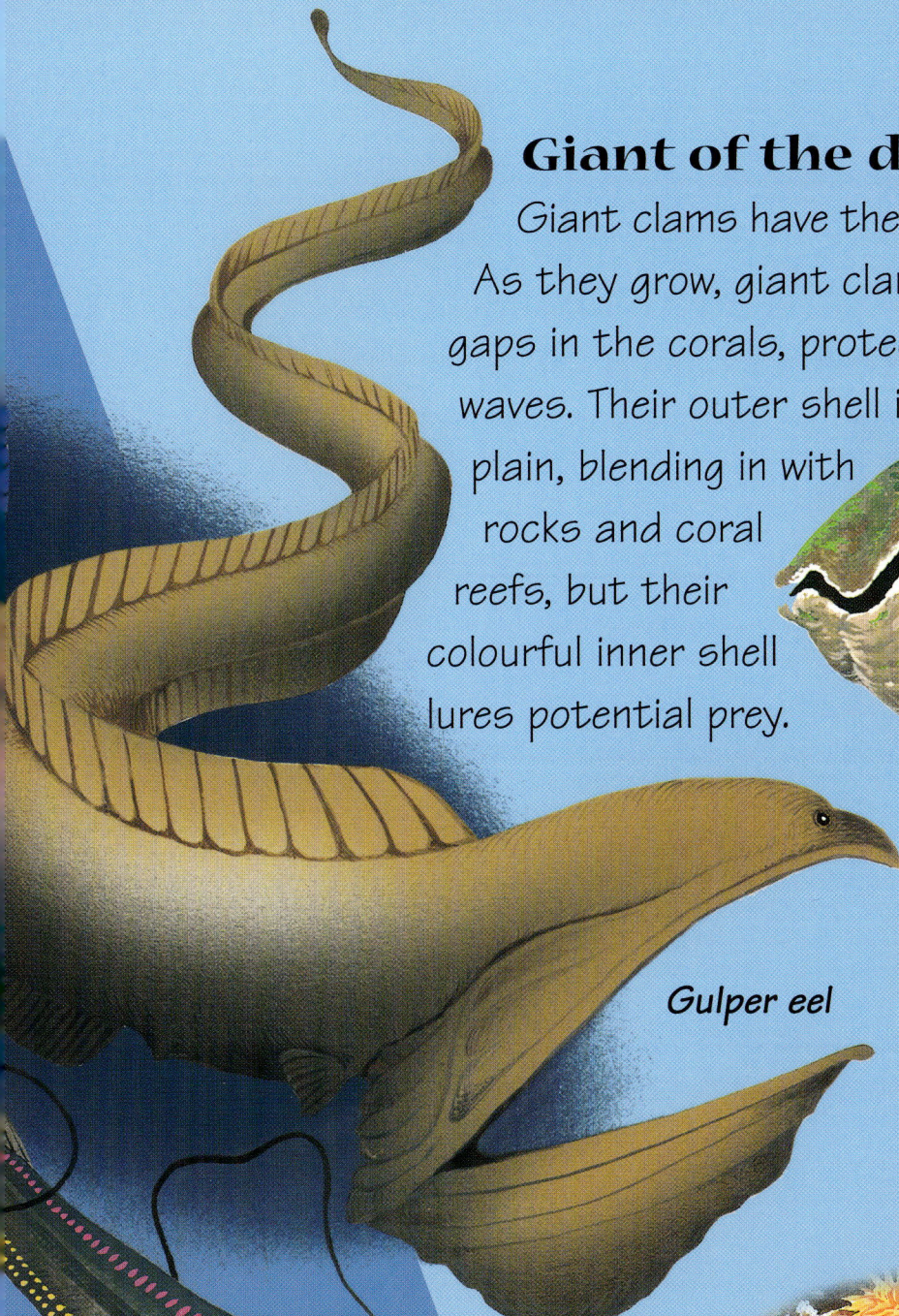

Gulper eel

TIGHT SQUEEZE
Moray eels live among coral reefs and can squeeze through small gaps between rocks at great speed to catch unsuspecting prey.

Dragonfish

In the bush

Woody™ has travelled across the world to Australasia. Because it is so hot in the day, many creatures are nocturnal, preferring to feed during the cool, fresh nights. Woody™ has met some red kangaroos, which are shy creatures. The males are red, but the females have greyer fur.

BIG BILLS

Kiwi birds use their long bills to feed on insects and other creatures that emerge when the soil is moist at night. They spend the day hiding, camouflaged by their brown feathers.

Wacky World

Kangaroos can grow as tall as a human adult and weigh up to 90 kilograms. But their babies are tiny when they are born – not much bigger than a cherry. The young, called a joey, crawls into a pouch on its mother's tummy, where it stays for about eight months.

WILD THING

Dingoes do not bark but instead howl like wolves. They hunt smaller mammals at night.

INTRUDERS

Possums are versatile, agile creatures. Although they live in woodlands, some work their way into homes and feed on scraps when no-one is looking!

Jungle japes

When Woody™ arrives in the jungle, all he can see are eyes shining out at him! Many nocturnal jungle animals are small animals related to monkeys, that live in the trees. They include mouse lemurs, whose wide, shiny eyes help them to see well in the dark. Lemurs feed on insects, fruit and flowers.

GOLDEN HUNTER

Pottos are silent, cautious hunters. They are hard to see in the jungle because of their golden fur. Their big, round eyes help them to see their prey at night and their round ears pick up tiny sounds, prompting them to freeze until danger has passed. They feed mainly on insects, particularly poisonous caterpillars.

HANGING ON

Slow lorises hunt at night for fruits and insects, and eat their food hanging upside down from tree branches so they can grip their meal with their front paws. Slow lorises are sensitive to light and sleep during the day.

22

LESSER BUSHBABY

These creatures have large ears like bats and huge dark eyes that take in light. They sleep in tree hollows during the day.

CLIMBERS

Tree pangolins are very good at climbing, and can climb their own scaly tails when in danger. They feed on termites.

FAKE OWL

Owl monkeys are the only members of the monkey family to hunt at night, feeding on small mammals, insects and fruit. They like to hunt during a full moon.

BEND IT

Tarsiers have brilliant night vision, a flexible neck that helps them to look for prey from all angles, and ears that move to trace any sound.

Africa at night

Woody™ is in the wilds of Africa and is looking forward to seeing the nocturnal animals there. He has spent the day watching the hippos. These creatures have to keep their oily skin wet, so they spend the hot days wallowing in water. At night, they graze, feeding mostly on riverside grass.

YOU CAN'T SEE ME!

Leopards wait for their prey on high tree branches, camouflaged by the light falling in spots on their coats. They stalk their prey from the ground, always seeking cover.

Wacky World

Nocturnal aye-ayes have a long, thin middle finger that helps them to dig insect larvae out of tree trunks and branches. Their huge, ears detect the tiniest movements.

Listen up

African bushbabies have an acute sense of hearing, and can find their insect prey very easily. Their enormous eyes help them to see in the dark. They have pads on their fingers and toes, helping them to cling to branches, and snatch prey from the air.

JUMP TO IT

Jerboas come out at night to feed on insects, plants and seeds. They use their long legs to leap out of danger like kangaroos.

FEATHERED FUN!
What do you give a seasick hippo?

Lots of room.

FEATHERED FUN!
On which day do leopards like to eat?

Chewsday !

Dry country

In the desert, nocturnal animals have to survive in a large, open area where they can be seen easily. They have to be able to blend in with their surroundings, to fend off attackers, or, like the jack rabbit, they have to be able to run — very, very fast! Woody™'s bright feathers would make it hard for him to hide in the open spaces.

COOL KILLER
Red-tailed hawks feed on small mammals, including bats. They can catch their prey while flying, using their talons to grip the unlucky victim.

THE BIG STING
Scorpions come out at night to hunt for small mammals and insects. They use a sting in their tails to kill them.

RATTLESNAKE
Rattlesnakes hunt at night when it is cooler. They warn attackers off by making a rattling noise with their tails.

HARE TODAY

Jack rabbits are actually hares. They live above ground and have much larger ears than rabbits. They can run at speeds of up to 72 kilometres an hour.

TINY DIET

Flammulated owls have a smaller appetite than many owls, feeding mostly on insects. They wait in a tree until they see their prey, then pounce on it.

SWIFT AND DEADLY

Coyotes live in the wilds of Northern America. They hunt by night, running at speed, guided by their excellent eyesight, hearing and sense of smell, and feed on live meat or dead animals.

All-seeing owls

Woody™ knows that most owls hunt during the night. But some owls, such as the burrowing owl, are also visible during the day. They live in underground caves and hunt on the ground as well as from perches, feeding on small mammals and insects.

SPECS APPEAL
Spectacled owls have white rings around their eyes. They live close to water and feed on birds, reptiles, insects and even bats! They are sometimes active during the day.

SILENT FLIGHT

Barn owls feed on small mammals, flying over fields to track them down. They nest in old farm buildings and sometimes church towers. They have unusual heart-shaped faces and no ear tufts.

FEATHERED FUN!

Why don't owls look for mates when it's raining?

Because it's 'too wet to woo'.

FEATHERED FUN!

Why did the owl 'owl?

Because the woodpecker would peck 'er.!

WHO DOES WHAT?!!

1. WHO HAS A SWEET TOOTH?
a) Cat
b) Silverfish
c) Slug
d) Dingo

2. WHO CLIMBS THEIR OWN TAIL?
a) Tree pangolin
b) Field mouse
c) Owl monkey
d) Rattlesnake

3. WHO IS A 'BANDIT OF THE NIGHT'?
a) Fox
b) Raccoon
c) Cat
d) Badger

4. WHAT ANIMAL LOOKS MORE LIKE A FLOWER?
a) Bat starfish
b) Giant clam
c) Fruit bat
d) Anemone

5. WHO GOES FISHING WITH A GLOWING ROD?
a) Toad
b) Glow-worm
c) Anglerfish
d) Hippo

6. WHICH OWL LIVES UNDERGROUND?

a) Spectacled owl
b) Burrowing owl
c) Barn owl
d) Owl monkey

9. WHICH INSECT FEEDS ON BLOOD?

a) Cockroach
b) Glow-worm
c) Giraffe weevil
d) Bed bug

7. WHO HAS A STING IN THEIR TAIL?

a) Possum
b) Jerboa
c) Scorpion
d) Hedgehog

10. WHO HAS A LONG, THIN MIDDLE FINGER?

a) Golden potto
b) Slow loris
c) Possum
d) Aye-aye

8. WHICH ANIMAL HOLDS BATTLES AT DUSK?

a) Earwig
b) Stag beetle
c) Cockroach
d) Sea horse

Answers
1. b; 2. a; 3. b; 4. d; 5. c; 6. b; 7. c; 8. b; 9. d; 10. d.

Index